Abel Hernández-Muñoz

Field guide to the bats of Cuba

Abel Hernández-Muñoz

Field guide to the bats of Cuba

A guide for children to identify species in nature

ScienciaScripts

Imprint
Any brand names and product names mentioned in this book are subject to trademark, brand or patent protection and are trademarks or registered trademarks of their respective holders. The use of brand names, product names, common names, trade names, product descriptions etc. even without a particular marking in this work is in no way to be construed to mean that such names may be regarded as unrestricted in respect of trademark and brand protection legislation and could thus be used by anyone.

Cover image: www.ingimage.com

This book is a translation from the original published under ISBN 978-620-2-15380-5.

Publisher:
Sciencia Scripts
is a trademark of
Dodo Books Indian Ocean Ltd. and OmniScriptum S.R.L publishing group

120 High Road, East Finchley, London, N2 9ED, United Kingdom
Str. Armeneasca 28/1, office 1, Chisinau MD-2012, Republic of Moldova, Europe
Printed at: see last page
ISBN: 978-620-7-28622-5

FIELD GUIDE TO THE BATS OF CUBA

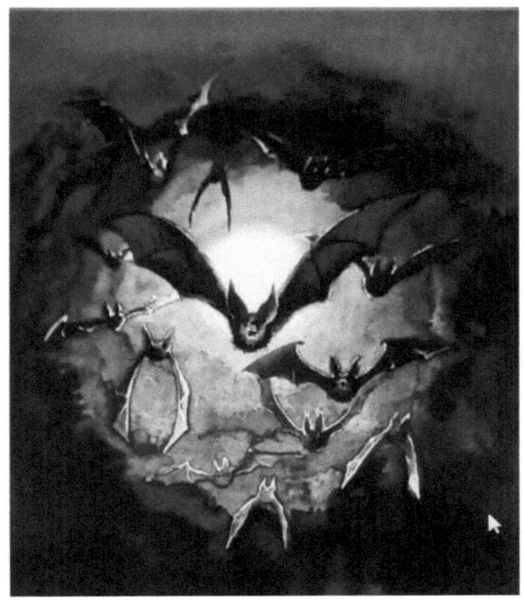

For children to identify species in nature.

MSc. Abel Hernández Muñoz

2023

FOREWORD

This book is an overview of our bats where common names, scientific names, descriptions, feeding, habitat, reproduction, distribution and conservation are detailed. It is profusely illustrated with full color drawings, as well as distribution maps in Cuba in black and white.

Field guides are a form of scientific popularization, already a classic in the world literature of the natural sciences, designed to provide amateurs with the means to quickly identify, under field conditions, a particular group of plants or animals in a given country or region. They are obviously characterized by an appreciable predominance of illustrations at the expense of text. Works of this type are in great demand internationally and their commercialization is usually very advantageous.

The identification and observation of bats in natural conditions is the object of a vigorous group of amateurs, among them the thriving international movement of fauna observers, who have been visiting us regularly for some years.

Its author is a specialist in fauna, chiropterologist of the Speleological Society of Cuba, has more than one hundred publications dedicated to the fauna of Cuba, which represents a modest contribution to the scientific knowledge of our vertebrates.

Author.

Table of Contents

INTRODUCTION

Bats are sensitive, vulnerable and little understood animals. Categorized as ugly and repulsive, they have been persecuted and harmed by some people intentionally and others due to lack of information.

Tons of bat droppings have been collected from the caves where they roost and applied to crops in the form of fertilizers have produced results similar to those obtained through the droppings of shorebirds.

This can be deduced from the examination of the layers that form the walls of these caves inhabited by bats that have been piled up over hundreds of years, some of them constituting deposits several meters thick.

The murcielaguine is very rich in phosphorus and exhales a pronounced ammonia odor that strongly attacks the eyesight.

Bats are also important agents in the pollination of plants, a biological phenomenon known as chiropterogamy.

Frugivorous species are responsible for disseminating the seeds of fruit and other trees, contributing to the renewal of the vegetation cover of forests and fields.

Bats perform another very important ecological function, that of biological controllers of insects, which are devoured by the ton with the help of these little animals, being allies of man by contributing to the fight against agricultural and forest pests or those that cause damage to human health and domestic animals. The occurrence of malaria in certain areas is related to the number of bats and varies in inverse proportion to the number of bats, the more bats, the fewer malaria-carrying mosquitoes.

Gray bugs devour up to three thousand insects in one night; in Texas, a colony eliminates two hundred and fifty thousand in one night; the Myotis consumes five hundred insects per hour, which is equal to one every seven seconds. Research reveals the increase of mosquitoes and other insects in proportion to the decrease or decline of the chiropteran population.

Cuban chiropterologist Gilberto Silva Taboada once said that every night, Cuban bats devour in our island no less than 150 tons of insects that are annoying to humans or harmful to agriculture. That means some 54 million kilograms a year. This is a very conservative calculation, estimated only for eleven of the twenty species of insectivorous bats that fly from dusk.

In other words, every year, 5 kilograms less insects per person, which will no longer harm ten million inhabitants.

These cute little animals have served as experimental material and in the development of vaccines. Their flight has been rigorously studied, contributing considerably to the development of aeronautics and airspace exploitation. Man designed and perfected radar based on their extraordinary echolocation system. The "sound images" that bats developed to perceive and capture their prey served as an antecedent to the sound scanner used by visually impaired people to orient themselves.

GENERAL CHARACTERISTICS OF THE ISLAND OF CUBA

Cuba, large island of the Caribbean Sea and the largest of the Antilles, located at the mouth or entrance of the Gulf of Mexico: it has 104 556 square kilometers of surface area. It is 1 250 km long from east to west from Cape S. Antonio to Punta de Maizi. Antonio to the Punta de Maizi, and 191 in its greatest width, and 31 where it is narrower: it has Florida and the Bahama Islands to the north, the island of Santo Domingo to the east, Jamaica to the south, and the Gulf of Mexico to the west: it is between 20^0 of north latitude where Cape Cruz is demarcated, and 23^0 and 15 minutes where the Bay of Matanzas is located, and from 88^0 3 minutes where Cape San Antonio is located, to 301^0 and 20 minutes from the Punta de Maisí. Its rivers, which are 148, abound with rich fish, its mountains of precious and thick woods of cedar, mahogany, oak, oak, granadillo, guaiac and ebony; the fields of game and song birds, flowering trees and odoriferous plants: the land is very fertile, so that never lack flowers in the field, nor the trees are bare of their leaves; there are many medicinal hot water baths in this island: It has 11 large and comfortable bays, very safe harbors and abundant salt flats. It is extremely abundant in sweet and bitter yuccas, which are used to make cassava bread, coffee, corn, indigo, cotton, some cocoa, cane sugar and lots of excellent quality tobacco.

The most particular thing about this climate is that at any time of the year the fields are delicious and so healthy that the sick go there to convalesce.

It is a biodiversity hotspot in the insular Caribbean. The flora is spectacular, with more than 7,000 species of which more than 50% are exclusive to the island. Its fauna is rich and with a high level of endemism. No beast or poisonous animal is found on the island.

CHARACTERISTICS OF THE CUBAN BAT FAUNA

From the point of view of classification, the order Chiroptera is divided into two major suborders: Megachiroptera and Microchiroptera. The former comprises the so-called flying foxes or fruit bats of Africa, Asia and Oceania. They are grouped in a single family and can reach a wingspan of 5 feet (152.4 cm), i.e. from wingtip to wingtip. They are therefore the giants of the order. They feed on plant matter, particularly fruits; they undertake large migrations, determined by the abundance or scarcity of food at different times of the year. They spend their resting hours in the trees and do not possess the ugly appearance of the members of the other suborder; their faces are elongated and free of excrescences or cutaneous appendages. Many of them lack the echolocation system. Characteristic is the position of a fingernail on the second finger of the hand. The second division includes the great majority of species. All the bats of Cuba belong to this division.

On our island there are twenty-six living species of bats, all grouped into twenty genera and these in turn into six families. According to studies carried out on fossil remains, six extinct species have been described, among which are the feared and well-known blood-sucking bats (vampires), a species known scientifically as *Desmodus puntajudensis.* Currently, these hematophagous bats are only found in Central and South America. All our species are harmless and highly beneficial, due to the enormous proportion of insects they consume and the large quantities of guano (fertilizer for plant cultivation) formed by the droppings of some species.

It is very common to observe large colonies of bats of sixteen species in the caves that are distributed throughout the national territory, they are also common in the ceilings of houses (especially in the countryside), another five species in palm trees, in the trunks of split and hollow palms, and in large and leafy trees; the remaining six species of Cuban chiroptera.

In summary, they have a wide distribution and use the resources of the environment for shelter and, consequently, with the purpose of guaranteeing their subsistence. Of the total number of living species, 19 are insectivorous, 4 are pollinivorous-nectarivorous, 2 are frugivorous and 1 is piscivorous. In addition, seven are threatened with extinction: *Pteronotus parnelli, Pteronotus quadridens, Pteronotus macleayi, Phyllonycterys poeyi, Mormopterus minutus, Antrozous koopmani and Natalus stramineus.*

The species reported as endemic to Cuba are the following: *Mormopterus minutus, Phyllops falcatus, Phyllonycteris poeyi, Natalus primus, Nycticeius cubanus, Lasiurus pfeifferi, Lasiurus insularis and Antrozous koopmani.*

This guide offers an idea of the great diversity of this order of mammals in the national territory. Logically this is only a sample, but you can appreciate the extraordinary variability of forms and specific anatomical details that undoubtedly make this group one of the most interesting of the Cuban fauna.

SPECIES DATA SHEETS

The following are the technical data sheets of the Cuban species of bats, but first we offer the reader a series of recommendations. For example, before determining the bat, it is necessary to observe it carefully, paying attention to the size of the body (mentally comparing it with the butterfly bat, the fruit bat and the fisher bat), the general tonality of the fur, as well as the presence of spots or stripes on the head, body and wings. The characteristic features of the bat's structure (shape of the face and tail, wingspan, length of the forearm, size of the ears, shape and size of the patagium, etc.), which are important diagnostic features for the determination of the species. It is also crucial to determine the trophic guild to which they belong and whether they fly in open spaces, among the forest canopy or in the undergrowth.

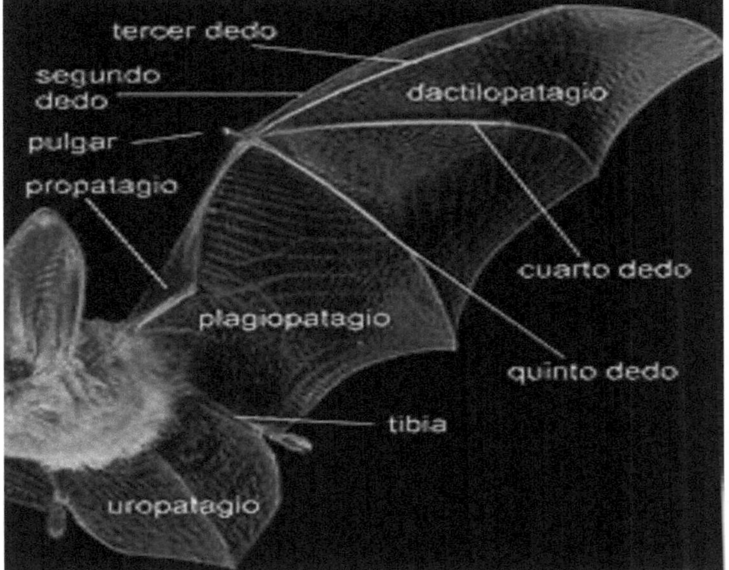

Fishing bat

(Noctilio leporinus)

ORDER: Chiroptera

FAMILY: Noctilionidae

Description. The fishing bat, as its common name indicates, includes fish in its diet, which it catches in flight with the large claws of its hind limbs. It is the largest chiropteran in Cuba, with a wingspan of 60 cm. It has a very characteristic yellow-orange color.

This chiropteran has been collected throughout the country. Its size is very large, with a wing span of sixty centimeters. The snout is short, wide and lacks a nasal blade. The lower lip has dermal folds. The ears are pointed and long. The male is larger than the female. The tail is proportional to the size of the animal, short and projects on the dorsal side of the center of the uropatagium. Its coat is short and its color ranges from dark brown to reddish brown on the back, with cream or orange tones in the ventral region. A light stripe stands out along this region.

Also known as:

Sexual dimorphism:

Wing expansion: 558-710 mm.

Weight: 54-87 g.

5
FISHERMAN BAT

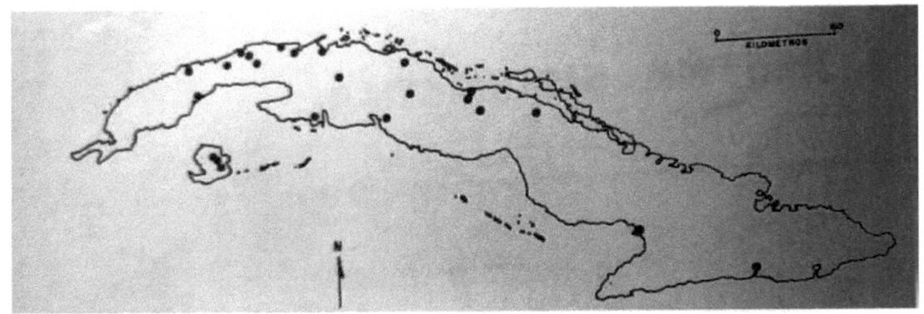
KILOMETROS
N

Mustached bat

(Pteronotus parnelli)

ORDER: Chiroptera.

FAMILY: Mormoopidae.

Description. Medium size. Short and rather broad muzzle. Without nasal flare, but with cutaneous excrescences above the nostrils; visible fleshy protuberance on the dorsum. Double transverse dermal fold in the middle of the lower lip. Moderately long and pointed ears set wide apart on the head.

Also known as:

Sexual dimorphism:

Length: 332-363 mm.

Weight: 9-15 g.

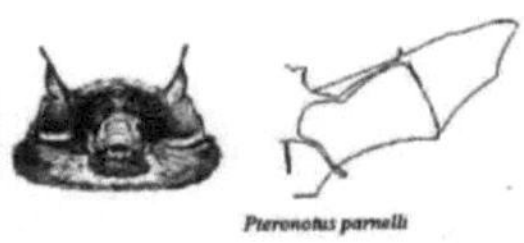

Pteronotus parnelli

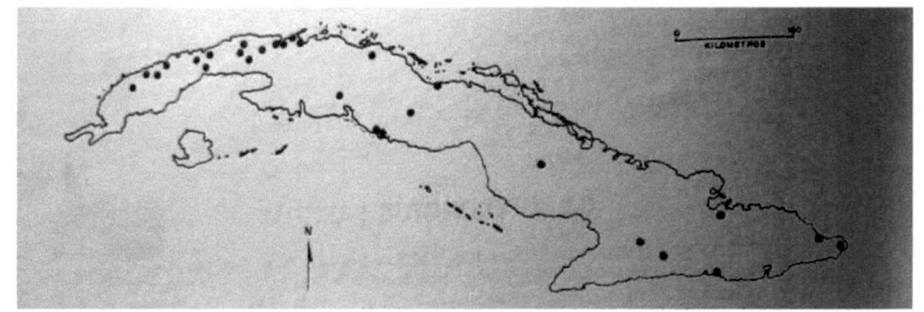

KILOMETROS
N

Mustached bat

(Pteronotus quadridens)

ORDER: Chiroptera.

FAMILY: Mormoopidae.

Description. Small size. Muzzle rather short; without a nasal blade, but with cutaneous excrescences above the nostrils and a fleshy tab on the sides of the nose. Double transverse dermal fold in the middle of the lower lip. Moderately long and pointed ears set wide apart on the head. Uropatagium very broad.

Also known as:

Sexual dimorphism:

Wing expansion: 229-276 mm.

Weight: 3-6 g.

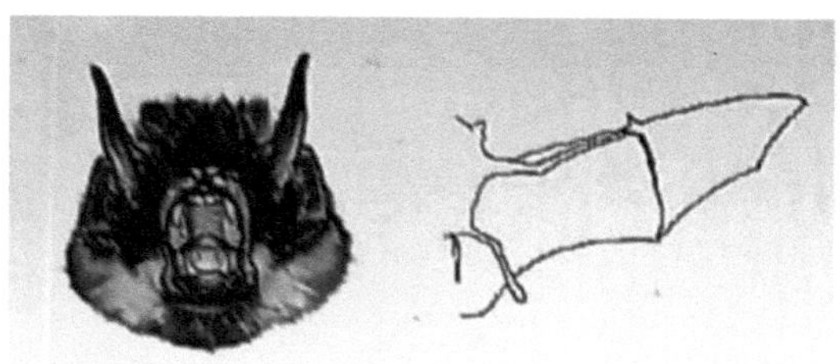

Pteronotus quadridens.

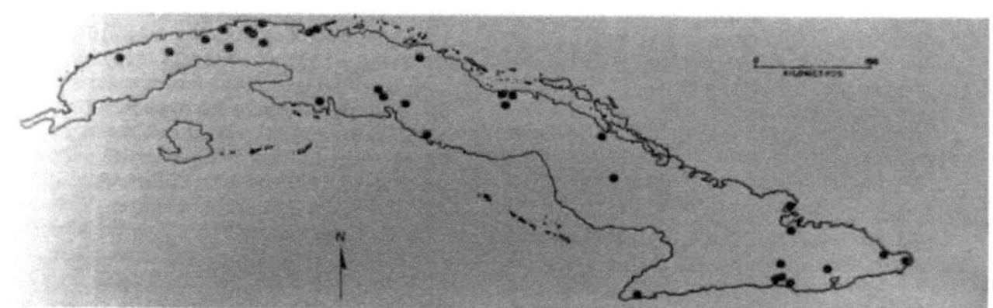

(Pteronotus macleayi)

ORDER: Chiroptera.

FAMILY: Mormoopidae.

Description. Medium size. Muzzle rather short; without a nasal blade, but with cutaneous excrescences above the nostrils and a fleshy tab on the sides of the nose. Double transverse dermal fold in the middle of the lower lip. Moderately long and pointed ears set wide apart on the head. Uropatagium very broad.

Also known as:

Sexual dimorphism:

Wing expansion: 269-296 mm.

Weight: 4-8 g.

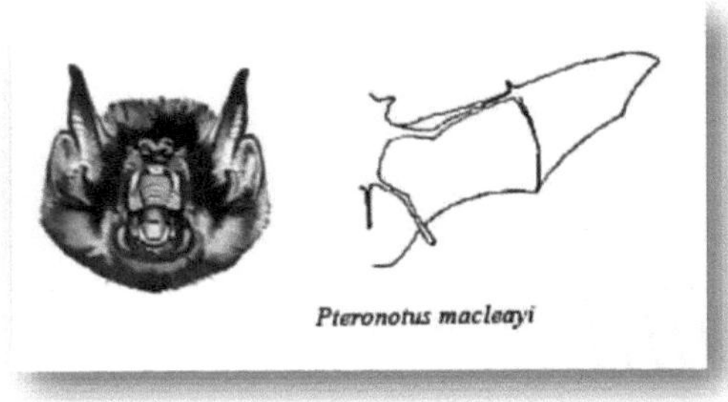

Pteronotus macleayi

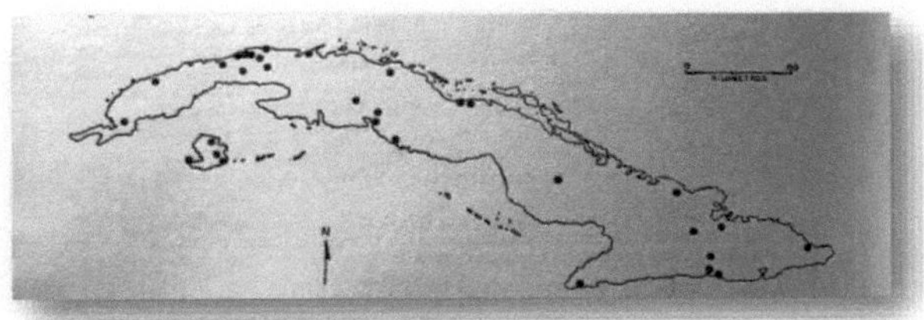

Northern Wrinkle-Bearded Bat

(Mormoops megalophylla)

ORDER: Chiroptera.

FAMILY: Mormoopidae.

Description. Northern wattled bats feed exclusively on large moths and are strong, fast fliers. They spend the day in caves or abandoned mine shafts and leave shortly after dusk to fly to streams and canyons where they forage for food. They return to the roost about seven hours later. A colony of northern wattled bats may have as many as half a million animals. Where several kinds of bats share a cave, they are kept separate from other species. These bats have small eyes and wrinkled lips that form a strange funnel-like structure. They also have a foliform jaw protrusion, which gives rise to two other common names: foliform jawed bat and old bat.

Also known as: Foliform Jawed Bat, Elderly Bat, Peter's Wrinkled Bearded Bat.

Sexual Dimorphism: None

Wing expansion: 78-98 mm.

Weight : 15-16 g.

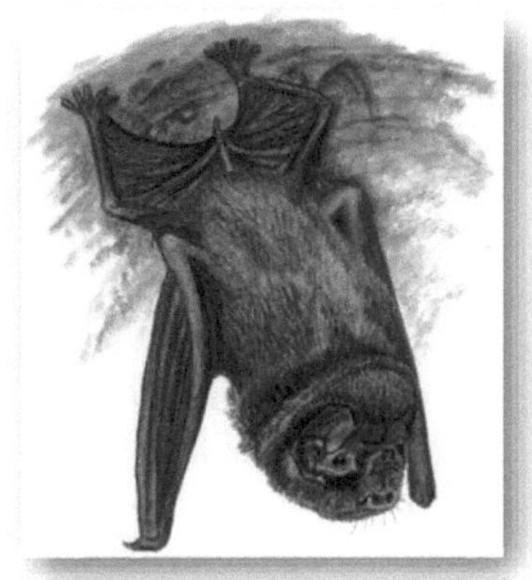

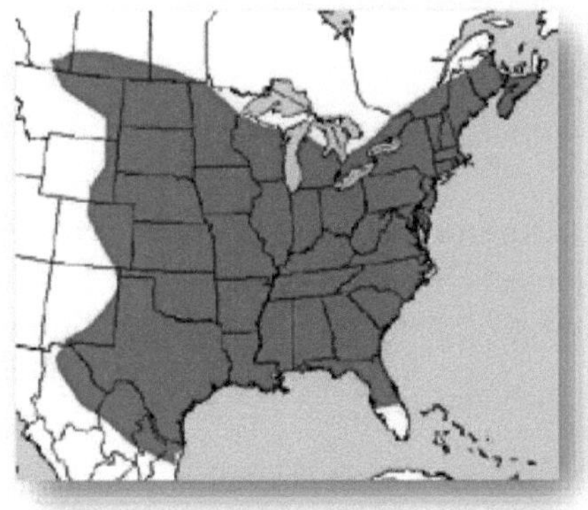

(Mormoops blainvillei)

ORDER: Chiroptera.

FAMILY: Mormoopidae.

Description. It is a medium-sized animal; with a very short muzzle, without a nasal blade, but with some excrescences on the upper edge of the nose; lower lip provided with complicated folds and folds, with a central plate in the form of a shield, covered with tiny warty tubercles. Ears very close to the forehead, and very open and short, so short that they barely protrude from the fur. Tail not very long, but projecting from the center of the distal side of the uropatagium. Uropatagium very broad. Spur very thin, but strong and long. Dense, soft and very long coat. The color varies from pale cinnamon to reddish cinnamon, with various shades in between. Large tuft of hair in the form of a moustache, between the nose and the corner of the mouth.

Extremely fast flight during the morning return to the refuge; not so dizzying when leaving but always faster and higher than in the other Mormopidae, and producing a loud buzzing sound very striking.

Also known as: Red Bat.

Sexual dimorphism:

Wing expansion: 330-335 mm.

Weight: 6-11 g.

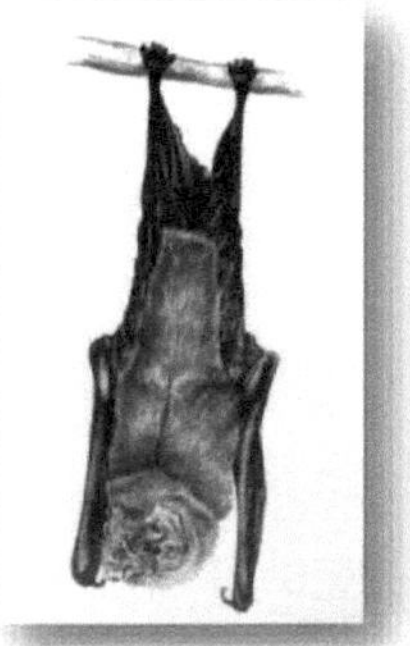

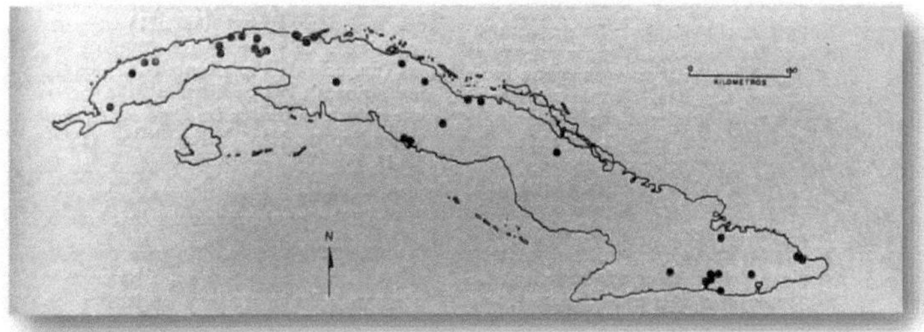
KILOMETROS
N

Eared Bat

(Macrotus waterhousei)

ORDER: Chiroptera

FAMILY: Phyllostomidae

Description. Eared bats usually employ their sense of sight (rather than echolocation) when foraging and resort to echolocation only in total darkness. They fly at a slow pace, close to the ground or vegetation and often catch butterflies and grasshoppers, which remain stationary at night when the bats hunt. They do not migrate or hibernate. They adapt to desert temperatures by seeking warm daytime roosts in caves, mines, or buildings. In winter, large groups of bats roost together in long, warm mine tunnels, usually in rocky areas heated by geothermal energy, and forage for only two hours each night. The young are born between May and July in maternity colonies that are often also located in caves. There are 100 to 200 females per maternity colony and each female has only one offspring.

Also known as: Guatacudo bat.

Wing expansion: 85-99 mm.

Weight: 12-22 g.

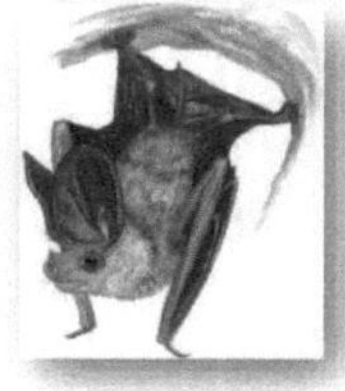 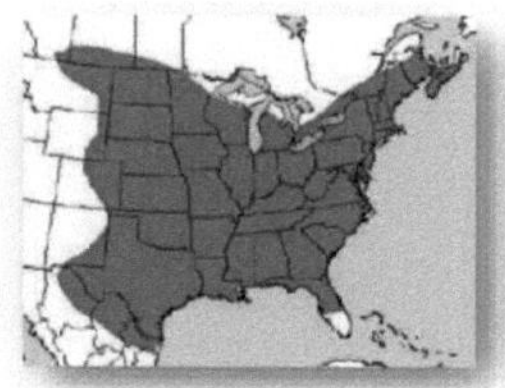 

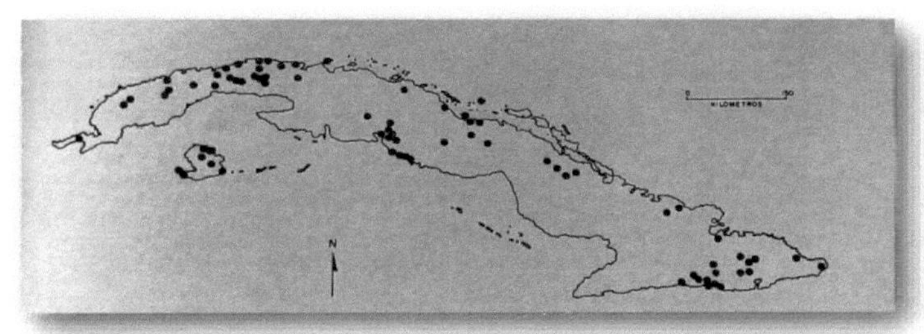

KILOMETROS
N

Long-tongued bat

(Brachyphylla nana)

ORDER: Chiroptera.

FAMILY: Phyllostomidae.

Description. Large size. Muzzle rather short and with dermal fold around the nostrils. Dermal lobes on the lower lip. Small but conspicuous wart behind the corner of the mouth. Ears proportionate and well separated on the head. Tragus lower in height than half of the ear.

Also known as: Screaming Bat.

Sexual dimorphism:

Wing expansion: 392-441 mm.

Weight: 27-41 g.

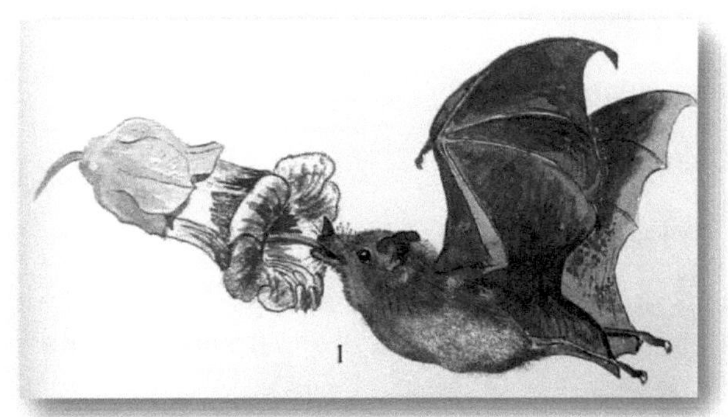

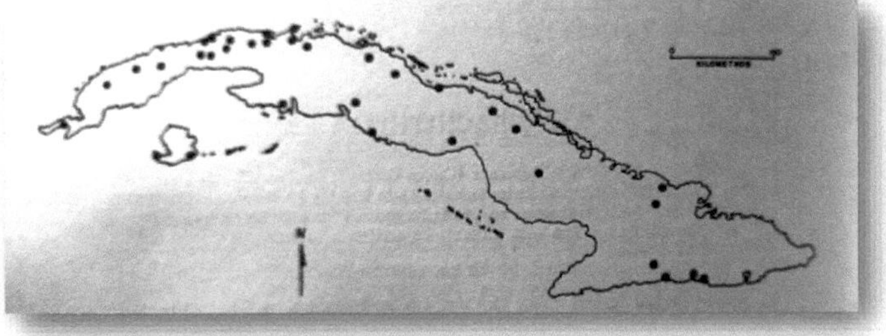

Flower Bat

(Erophylla sezekorni)

ORDER: Chiroptera.

FAMILY: Phyllostomidae.

Description. Medium size. Elongated and thin muzzle. Lower lip cleft centrally. Ears proportionate and well separated on the head.

Also known as:

Sexual dimorphism:

Wing expansion: 303-340 mm.

Weight: 13-21 g.

CUBAN FLOWER BAT

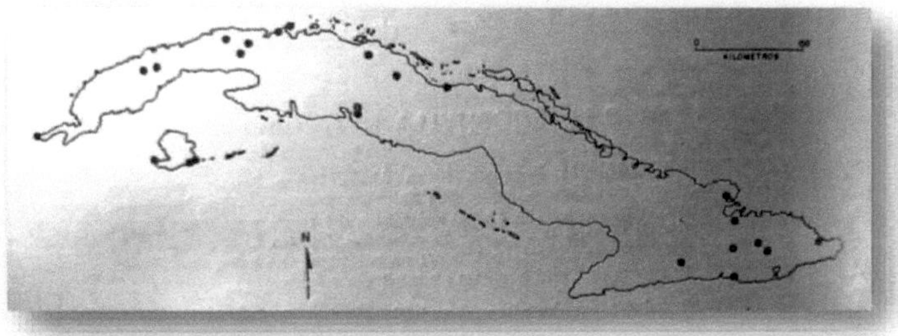
KILOMETROS
N

Hot Cave Bat

(Phyllonycteris poeyi)

ORDER: Chiroptera.

FAMILY: Phyllostomidae.

Description. Medium size. Muzzle elongated and thin, with dermal fold around the nostrils. Lower lip cleft centrally. Ears proportionate and well separated on the head. Distal end of the muzzle lower than the middle of the ear.

Also known as: Poey's bat.

Sexual dimorphism:

Wing expansion: 294-350 mm.

Weight: 15-29 g.

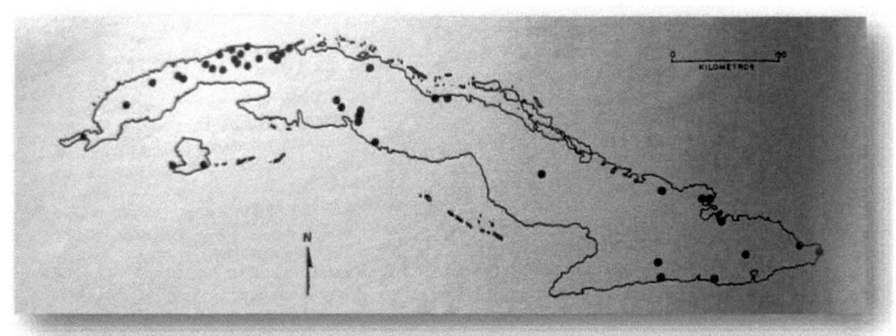

KILOMETROS
0   90
N

Nectarivorous Bat

(Monophyllus redmani)

ORDER: Chiroptera.

FAMILY: Phyllostomidae.

Description. Small size. Very long and thin muzzle, with prominent and pointed nostrils. Long tongue. Ears proportionate and well separated on the head. Small swallow.

Also known as:

Sexual dimorphism:

Wing expansion: 273-304 mm.

Weight: 8-14 g.

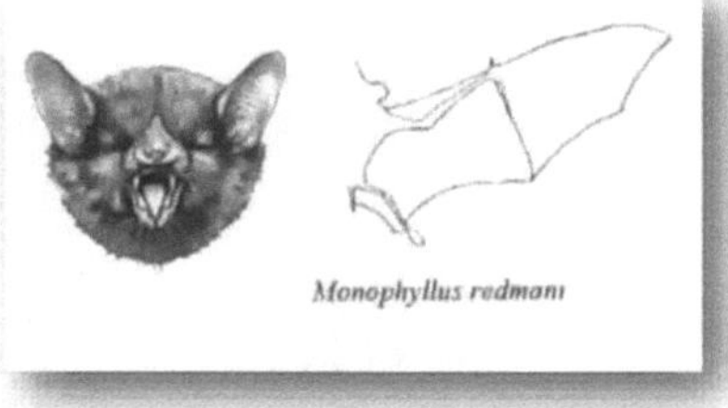

Monophyllus redmani

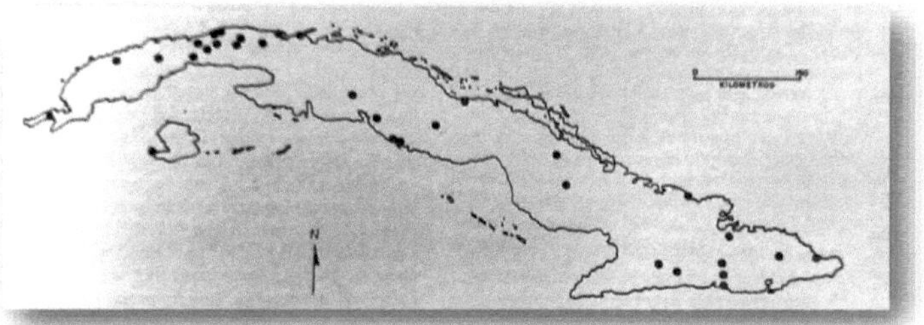

Fruit Bat

(Artibeus jamaicensis)

ORDER: Chiroptera.

FAMILY: Phyllostomidae.

Description. Large bat. Short and wide snout, with prominent and pointed nose blade. Lower lip with dermal lobules. Ears proportionate and well separated on the head. Large muzzle, but lower than the middle of the ear.

Also known as:

Sexual dimorphism:

Wing expansion: 340-445 mm.

Weight: 27-45 g.

 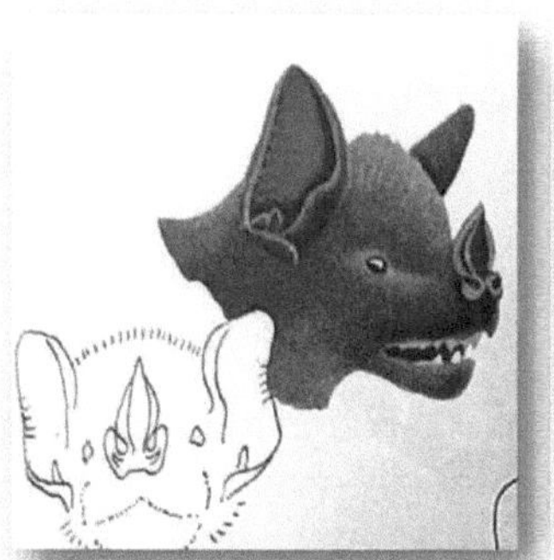

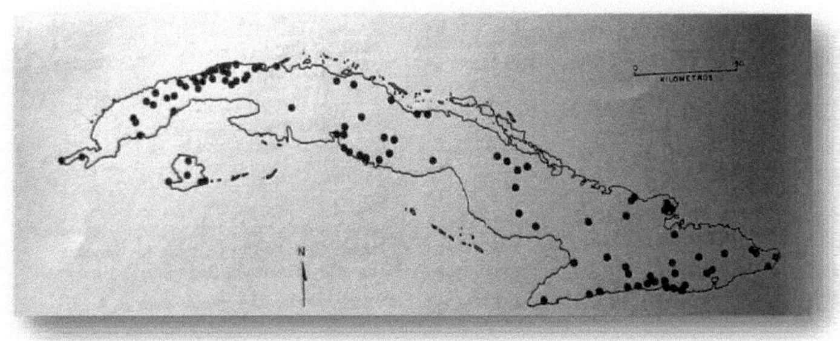

Little Fruit Bat

(Phyllops falcatus)

ORDER: Chiroptera.

FAMILY: Phyllostomidae.

Description. Medium-sized chiropteran. Very short and wide snout, with prominent and pointed nostrils. Ears proportionate and well separated on the head. Tail absent. Uropatagium reduced. Very short dewclaw. Dense and not very short coat. General color, brownish gray; darker on the back. Feeds mainly on fruits. Roosts in the foliage of broad-leaved trees such as Honduras mahogany. This bat seems ready to penetrate at night in the country houses. It is one of the three endemic species of Cuba.

Also known as:

Sexual dimorphism: female larger than male.

Wing expansion: 315-365 mm.

Weight: 16-23 g.

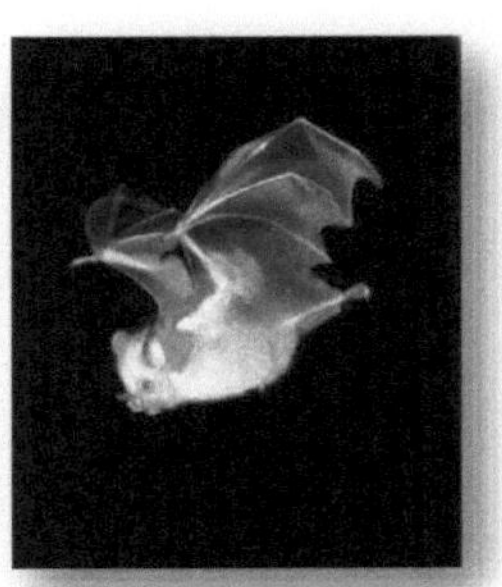

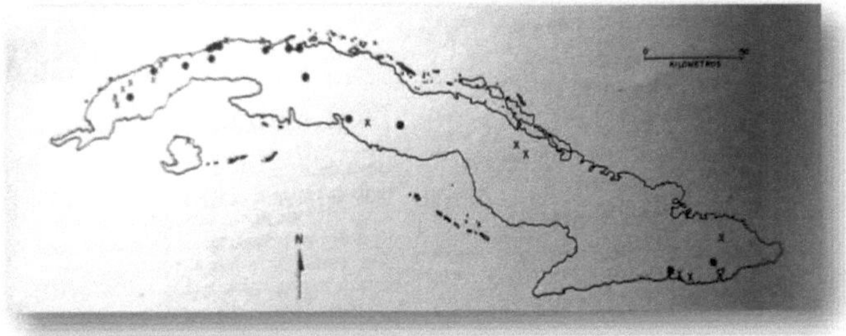

KILOMETROS
N

Butterfly Bat

(Nytielus lepidus)

ORDER: Chiroptera

FAMILY: Natalidae.

Description. Extremely small. Muzzle wide and short; without nasal blades or skin excrescences. Middle zone of the lower lip, thickened. Ears proportionate and separated on the head. Swallow at the height of the middle of the ear. Its distinctive feature is that it is the smallest of all bats living in Cuba and one of the smallest mammals in the world. It usually flies very low, almost at ground level. It lives in caves and feeds exclusively on insects, mainly mosquitoes, making it very useful to man. Its flight is erratic like that of a butterfly, reaching a much smaller size than many of them.

Also known as:

Sexual dimorphism:

Wing expansion: 186-213 g.

Weight: 2-3 g.

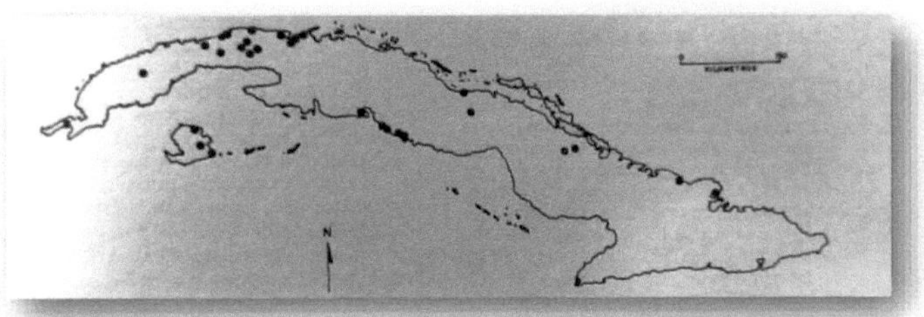

KILOMETROS
N

Large Funnel-Eared Bat

(Natalus primus)

ORDER: Chiroptera.

FAMILY: Natalidae.

Description. Medium-sized bat. It is the largest of the Cuban natálidos. Long, narrow and dorsoventrally depressed snout. Remarkable fleshy swelling (natalid organ) in the male, between the ears. Ears approximately as wide as high, have square shape with rounded tip and are relatively long, separated on the head. This species is endemic to Cuba.

Also known as:

Sexual dimorphism:

Wing expansion: 243-252.

Weight: 7-13 g.

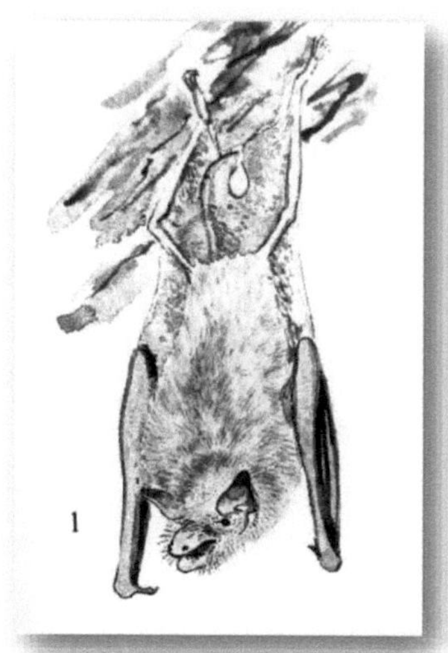

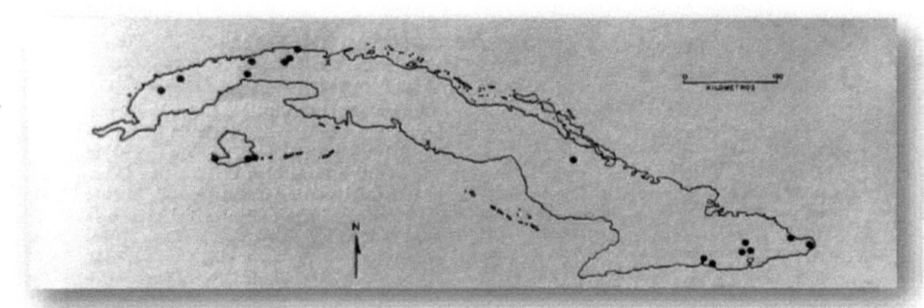

Little Funnel-eared Bat

(Chilonatalus macer)

ORDER: Chiroptera.

FAMILY: Natalidae.

Description. Very small. Narrow and depressed snout from the dorso-ventral perspective. Remarkable fleshy swelling (natal organ) in the male, between the inner bases of the ears. Tiny fleshy protuberance, somewhat projected forward and above the nose. Ears approximately as wide as high, set wide apart on the head. Small and twisted swallow.

Also known as:

Sexual dimorphism:

Wing expansion: 223-242 mm.

Weight: 2-4 g.

Carmelite Bat

(Eptesicus fuscus)

ORDER: Chiroptera

FAMILY: Vespertilionidae

Description. Carmelite bats live in rural areas, towns and cities and sometimes choose barns, houses or other buildings as roosts. Males usually live alone; females gather in maternity colonies during the spring and summer for the birth and care of their young. A maternity colony may have as many as 20 to 75 adult females and their offspring. Females in the eastern United States usually have twins; those in the west usually have only one offspring per year. Females may return to the same colony year after year. On warm, dry nights, bats leave the roost shortly after sunset in search of insects, especially flying beetles, which they capture and eat in the air. When the weather is cold or wet, they may stay in the shelter; their body temperature drops and they live on stored fat. In the winter, they hibernate. Many migrate short distances (less than 80 kilometers) in search of mines or caves to hibernate, but some spend the winter in attics or walls where the temperature is cold but remains above freezing.

Also known as: North American Brown Bat, Big Brown Bat.

Sexual Dimorphism: Females are larger than males.

Wing expansion: 87-138 mm

Weight: 16 -11g.

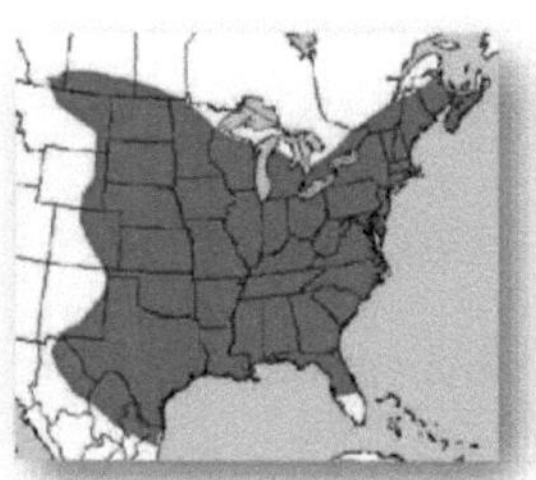

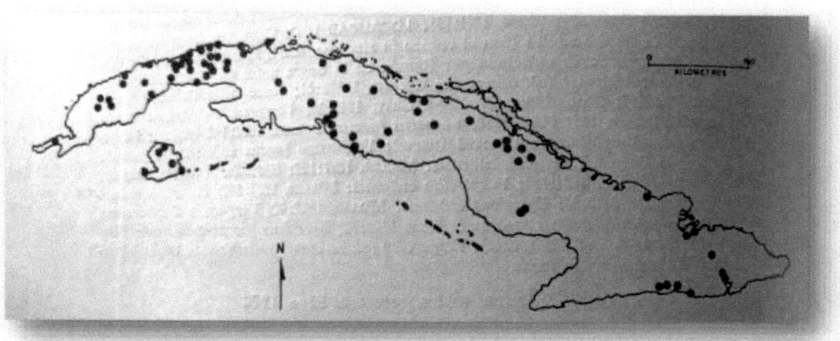

KILOMETROS
N

American Twilight Bat

(Nycticeius humeralis)

ORDER :Chiroptera

FAMILY :Vespertilionidae

Description. American crepuscular bats that roost together appear to share information about the location of rich foraging areas and alternate roosting sites. When foraging on farms, they are of great benefit to farmers because they eat cucumber beetles (adult stage in the life cycle of the southern corn rootworm). American crepuscular bats have never been found sheltering in caves. Summer maternity colonies have been discovered in buildings and hollow trees; in the winter, bats of this species have been found, in Florida, sheltering in the leaves of palm trees. Females and young appear to migrate quite large distances; one bat was located 520 kilometers from where it was captured and tagged. Only females migrate northward in the summer; males apparently remain in warm southern locations throughout the year.

Sexual Dimorphism: Females are larger than males.

Wing expansion: 83-96 mm

Weight : 9-14 g.

MURCIÉLAGO DEL
CREPÚSCULO
Nycticeius cubanus *

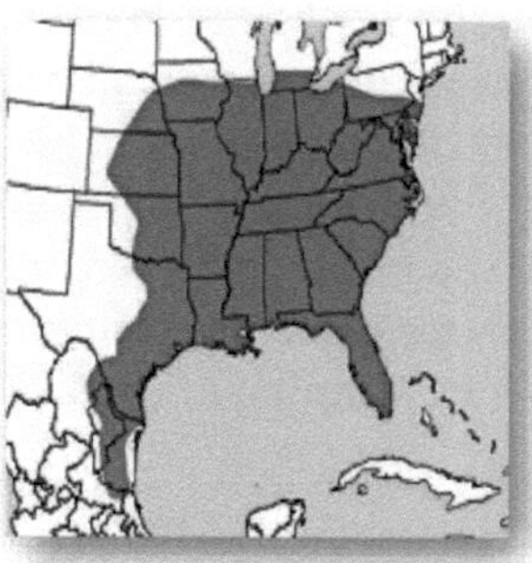

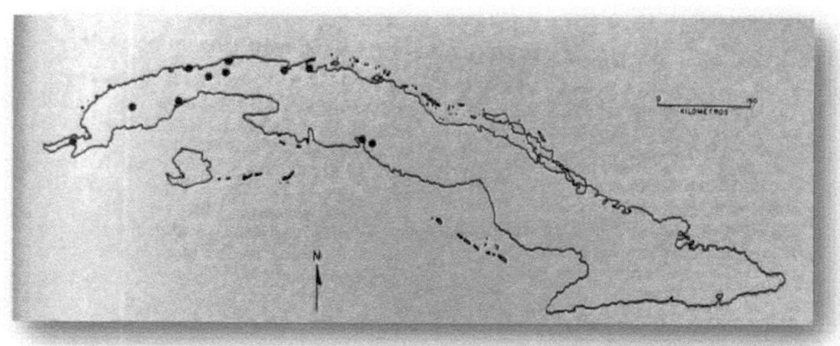

Cuban Twilight Bat

(Nycticeius cubanus)

ORDER: Chiroptera.

FAMILY: Vespertilionidae.

Description. This bat is endemic to Cuba. It is common in localities of the western region, especially in Havana. So far it has not been recorded for the eastern region of the island. It is a small-sized bat. It is the smallest vespertilionid in Cuba. Its snout is short and its fur is dense and brownish.

Sexual Dimorphism: Females are larger than males.

Wing expansion: 75-105 mm.

Weight: 5-7 g.

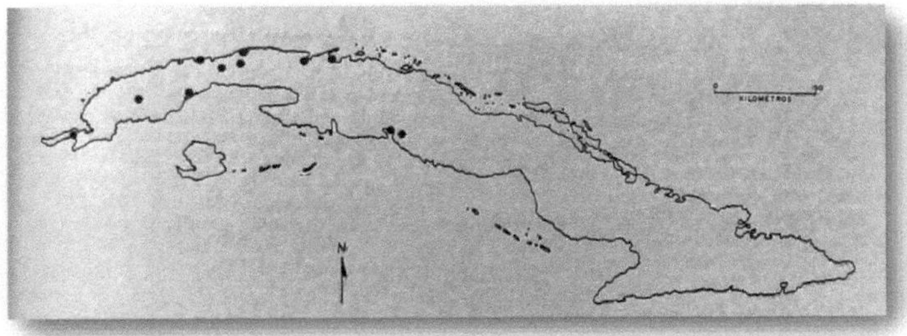

Reddish Hairy-tailed Bat

(Lasiurus pfeifferi)

ORDER: Chiroptera.

FAMILY: Vespertilionidae.

Description. Medium size. Muzzle broad and short, and somewhat bulging laterally, between the eye and the nostril; without nasal flake or skin excrescences. Ears wide and short, and separated on the head. Distal end of the tragus half the height of the ear.

Also known as: Pffeifer's bat.

Sexual dimorphism:

Wing expansion: 330-360 mm.

Weight: 8-14 g.

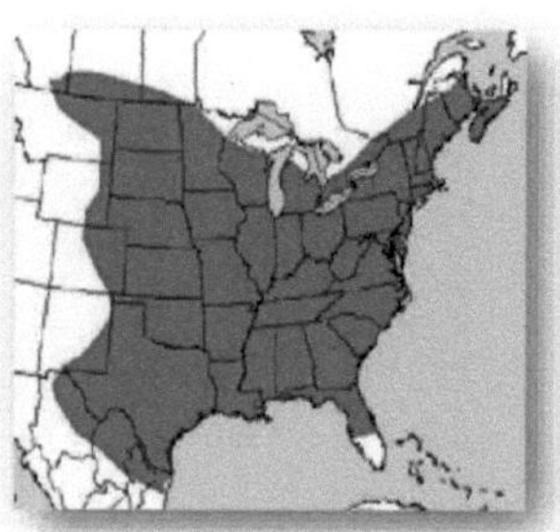

Lasiurus pfeifferi - **female (left) and male (right).**

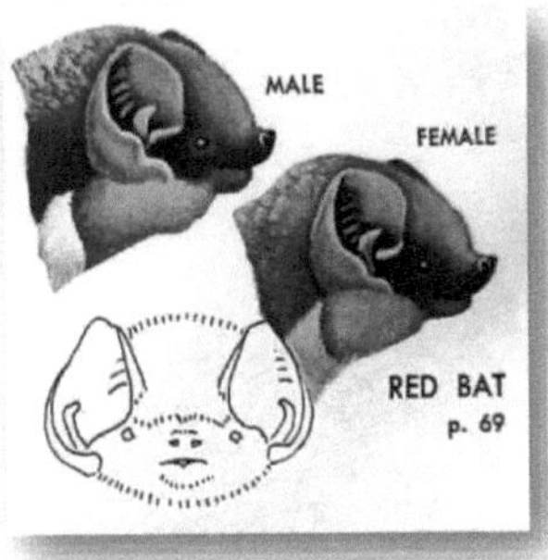

MALE
FEMALE
RED BAT
p. 69

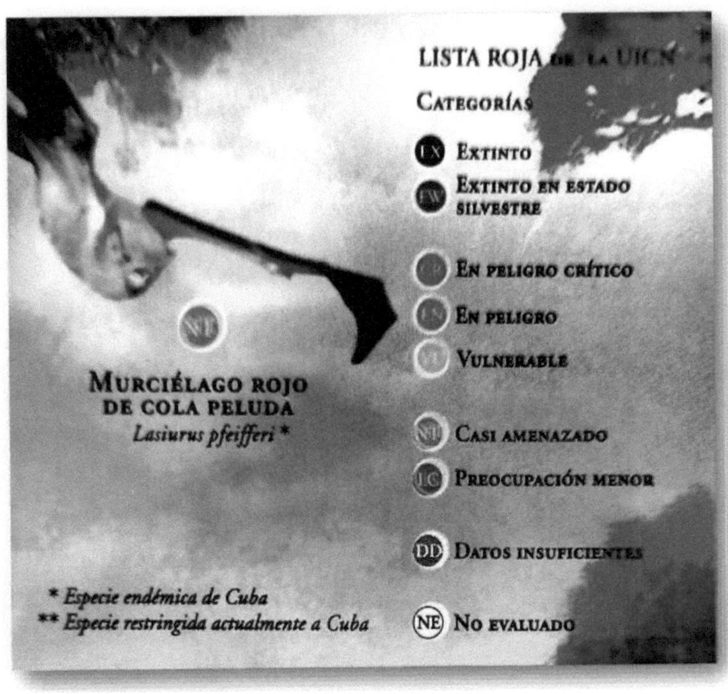

LISTA ROJA DE LA UICN
CATEGORÍAS
EX EXTINTO
EW EXTINTO EN ESTADO SILVESTRE
EN PELIGRO CRÍTICO
EN PELIGRO
VULNERABLE
NT CASI AMENAZADO
LC PREOCUPACIÓN MENOR
DD DATOS INSUFICIENTES
NE NO EVALUADO
MURCIÉLAGO ROJO DE COLA PELUDA
Lasiurus pfeifferi *
* Especie endémica de Cuba
** Especie restringida actualmente a Cuba

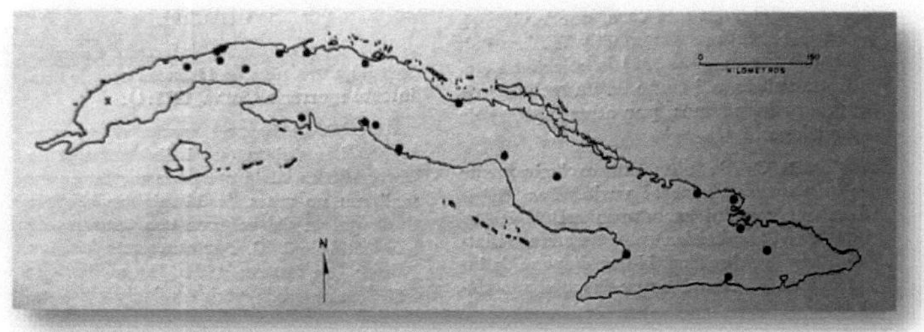

N
KILOMETROS

Big Hairy-tailed Bat

(Lasiurus insularis)

ORDER: Chiroptera.

FAMILY: Vespertilionidae.

Description. It is endemic to Cuba. It is one of the rarest bats on the island because it inhabits less than ten localities distributed throughout the country and has not been captured in the last 50 years. Due to its rarity and habits, it is considered vulnerable to extinction. It is a medium-sized bat. Its snout is wide and short, without facial ornamentation. Its ears are long.

Also known as: Yellow Bat.

Sexual dimorphism:

Wing expansion: 420-468.

Weight: 20-30 g.

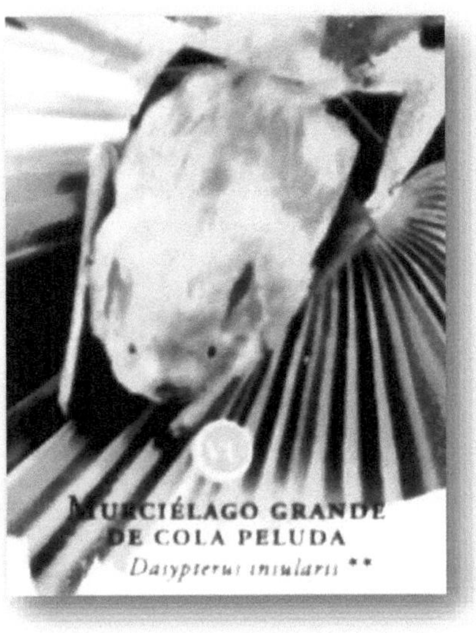

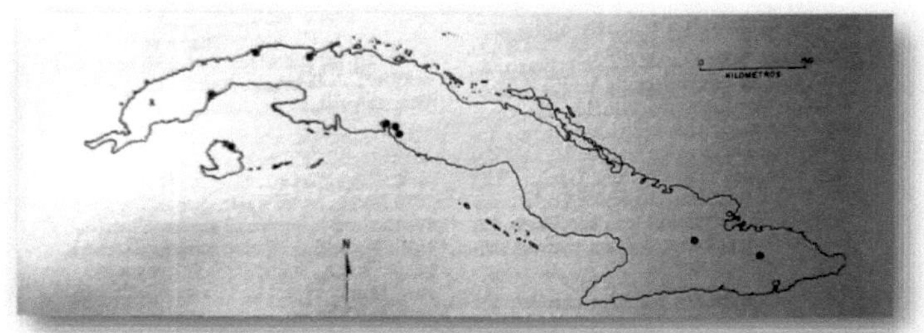

N
KILOMETROS

Koopman's bat

(Antrozoos koopmani)

ORDER: Chiroptera.

FAMILY: Vespertilionidae.

Description. It is an endemic bat of Cuba, very rare, which for this reason is considered Vulnerable to extinction. It is of medium size. Its snout is wide and short, without facial ornamentation. Large ears of more than 20 mm.

Also known as:

Sexual dimorphism:

Wing expansion: more than 380 mm.

Weight: Unknown.

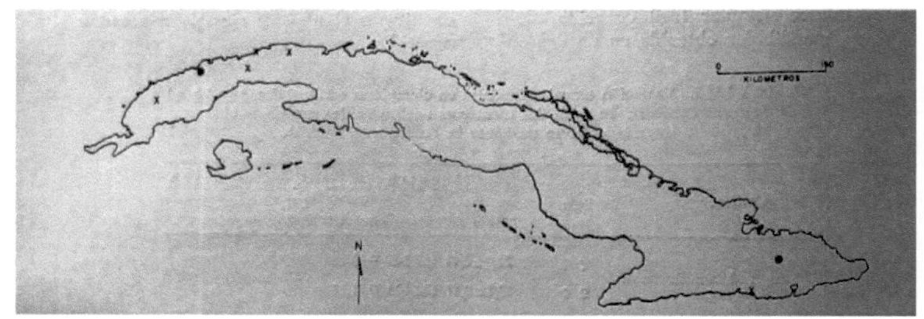

KILOMETROS
N

Pale Bat

(Antrozous pallidus)

ORDER :Chiroptera

FAMILY: Vespertilionidae

Description. The northern desert bat, common throughout its range, is found in arid and semi-arid regions throughout northern Mexico and the western United States. It eats beetles, grasshoppers, and moths and searches for slow-moving prey such as scorpions, flightless arthropods, and sometimes lizards on or near the ground. It uses echolocation to detect prey, but also its large ears to listen for prey movements. These bats visit flowers in their search for insects and are natural pollinators of several species of cacti.

Also known as: Northern Desert Bat.

Wing expansion: 92-135 mm.

Weight : males 13.6-24.1 g; females 13.9-28.9 g.

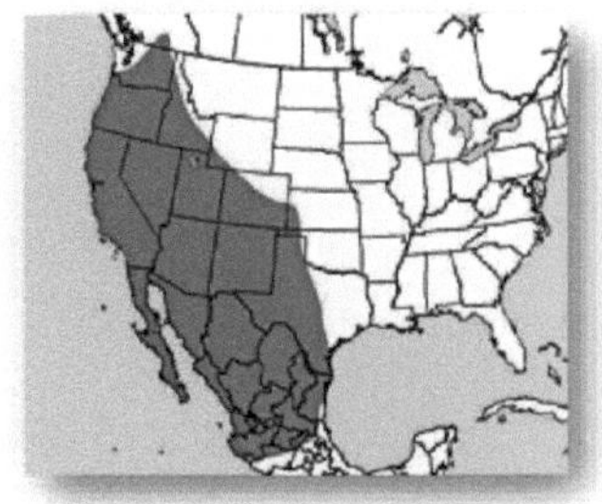

Guanero Bat

(Tadarida brasiliensis)

ORDER: Chiroptera

FAMILY: Molossidae

Description. Millions of Brazilian free-tailed bats spend the summer in the southwestern United States. Giant colonies summer in Bracken Cave, Texas; the caverns of Carlsbad, New Mexico; and even in the city of Austin, Texas, under the Congress Avenue Bridge. They offer a spectacular sight, spiraling out of their daytime roosts like a swirl of great dark clouds in the early evening in search of food. Bats eat countless insects each night and sometimes hunt their prey at heights of a mile or more. They typically migrate to the central and southern regions of Mexico in the winter, where they live in small colonies. They mate allrí and fly north again - up to 1,300 kilometers - between February and April. Females have a single brood in June and feed it for about six weeks. Although their numbers are in the millions, their conservation status is of concern because they raise newborns in a limited number of caves and because pesticides can accumulate in their body tissues.

Also known as: Mexican Loose-tailed Bat.

Sexual Dimorphism: Males may be 5% longer than females, but females weigh about 5% more than males.

Wing expansion: 85-109 mm.

Weight: 10-15 g.

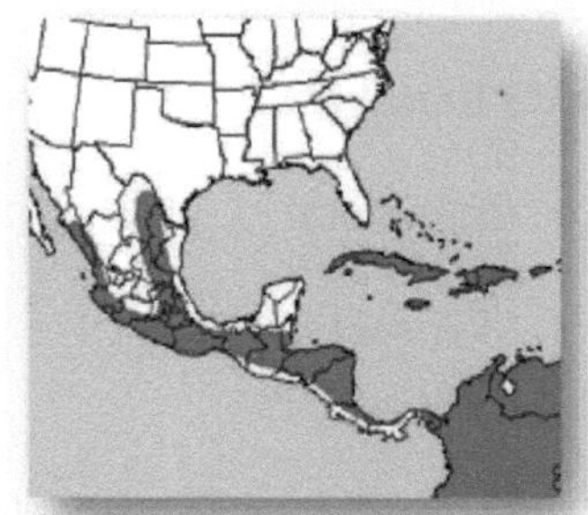

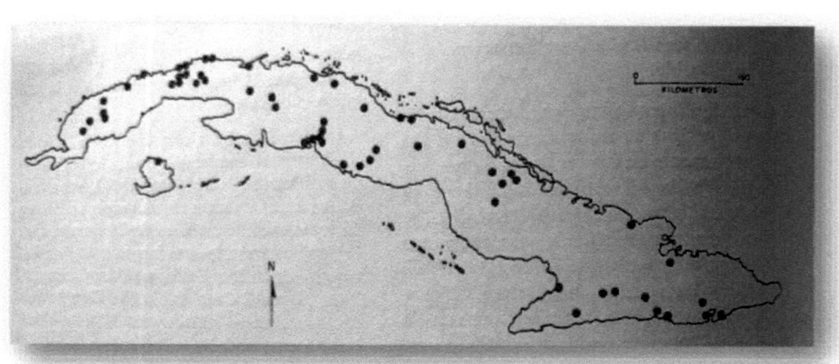

Whiptail Bat

(Nyctinomops laticaudatus)

ORDER:

FAMILY:

Description. Medium size, with a narrow muzzle, without nasal flares. Very wrinkled upper lip. Ears as wide as long and united on the forehead; with a remarkable longitudinal fin or keel on the ventral side, projected over the eye. Tiny swallow.

Also known as:

Sexual dimorphism:

Wing expansion: 305-335 mm.

Weight: 8-13 g.

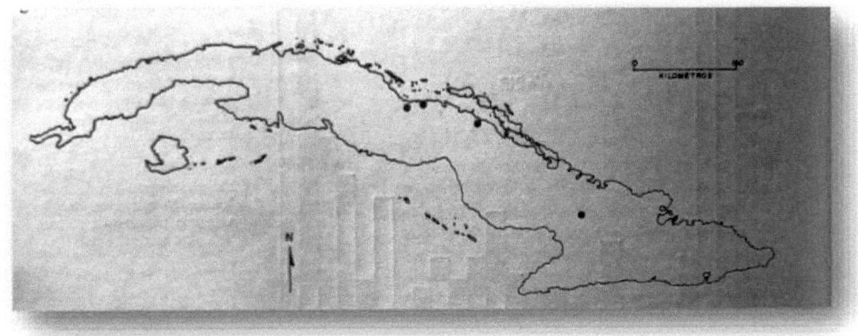

Bat with Greater Loose Tail

(Nyctinomops macrotis)

ORDER: Chiroptera

FAMILY: Molossidae

Description. The greater free-tailed bat has long, narrow, tapered wings. The length and shape of the wings give it great speed and enable it to fly long distances, but its flight does not allow it to maneuver as much as bats with shorter, broader wings. These bats live in the rugged habitats of the southwestern United States in the summer and migrate to Mexico in the winter. When foraging, they emit echolocation calls that sound like a clicking sound to the human ear. Almost all of these bats emit sounds that are beyond the threshold of human hearing. The bats forage for food, mostly large nocturnal butterflies, in complete darkness, and do not leave their daytime roosts until well after sunset. The tail extends beyond the caudal membrane (uropatagium), which stretches between the hind legs.

Sexual Dimorphism: Males are slightly larger than females.

Wing expansion: males 145-160 mm; females 120-139 mm.

Weight: 22-30 g

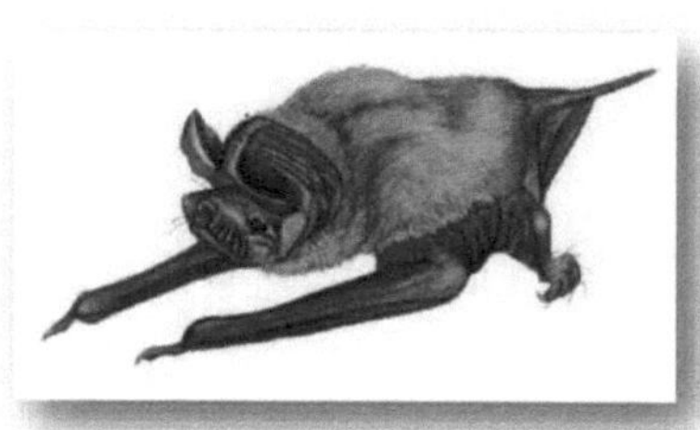 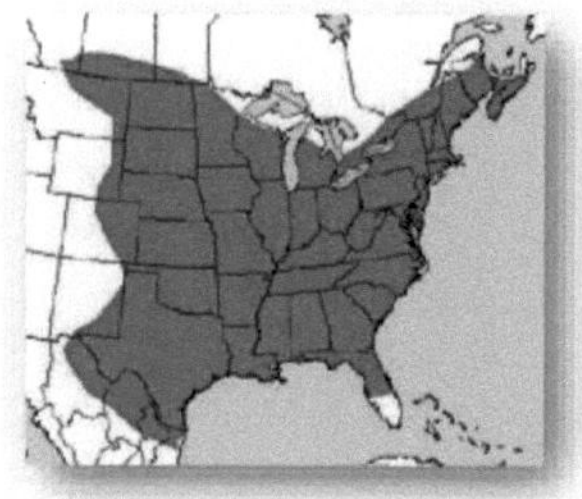

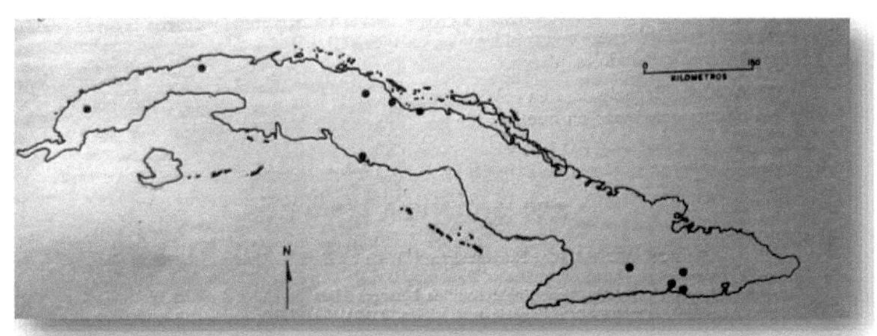

N
0   KILOMETROS   100

Bat of the Jatas

(Mormoopterus minutus)

ORDER: Chiroptera.

FAMILY: Molossidae.

Description. Small bat; elongated and slender snout, without nasal blades. Upper lip protruding considerably from the lower lip, almost twice as long as the lower lip. Ears proportionate, and somewhat close to the forehead. Tail very long and protruding considerably over the free edge of the uropatagium. Uropatagium moderately broad. Dense and short coat; brownish gray. Its natural shelter seems to be the palm *Copernicia vespertilionum,* commonly known in Cuba as "jata los murciélagos". Generally, this palm reaches in adulthood a height of twelve to fourteen meters, with a trunk between thirty and forty centimeters in diameter. The foliage is globular and compact; its upper half is formed by erect green tufts, while the lower half (where the bats settle) is a dense mass of dry and rigid tufts, very compact. The palms occupied by these bats are easily identified by the accumulation of guano around the base of the trunk at a height of up to sixty centimeters, although sometimes the guano is collected by farmers for use as organic fertilizer. This species is exclusive to Cuba, so special attention should be paid to it.

Also known as: **Little Bat of the Jatas**

Sexual dimorphism:

Wing expansion: 204-233 mm.

Weight: 4-8 g.

Murciélago chico
de las Jatas
Mormopterus minutus *

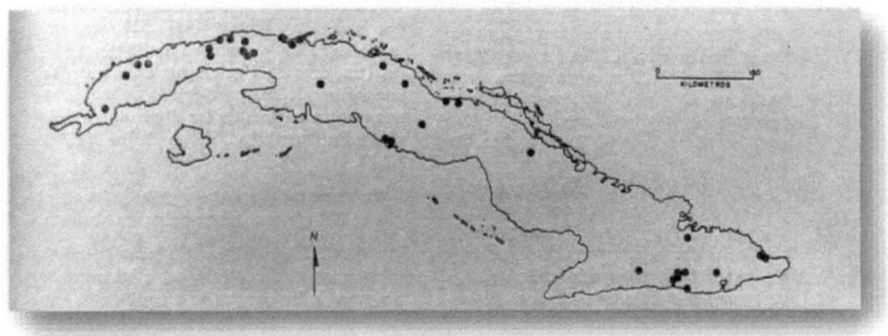

KILOMETROS
N

Wagner's Bat with Bonnet

(Eumops glaucinus)

ORDER :Chiroptera

FAMILY :Molossidae

Description. Wagner's bonnet-nosed bat is considered an endangered species in the state of Florida, where habitat destruction and pesticide use may contribute to its decline. It is a medium-sized bat with long, narrow wings. These bats leave their daytime roosts after dark and fly high, covering great distances quickly as they feed on insects. *Eumops glaucinus* bats are loosely tailed. Like other bats in the family Molossidae, the tail extends beyond the caudal membrane or uropatagium (the skin that stretches between the hind legs).

Also known as: Bat with Bonete de La Florida.

Wing expansion: males 123-165 mm; females 117-156 mm.

Weight: males 25-47 g; females 28.2-55.4 g.

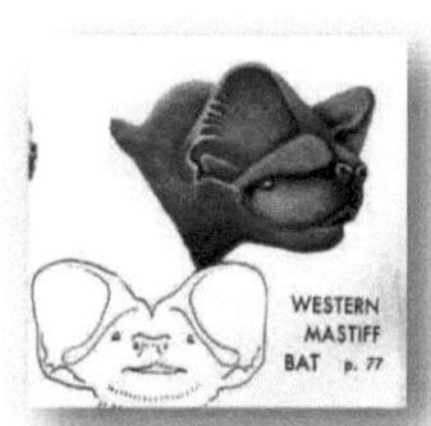

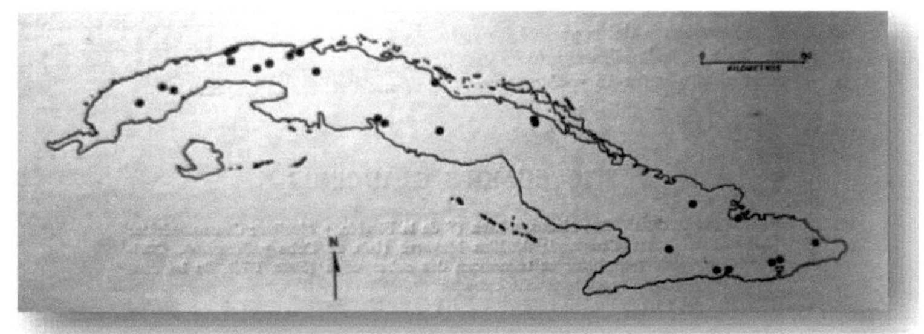

Homemade Bat

(Molossus molossus)

ORDER :Chiroptera

FAMILY :Molossidae

Description. This loose-tailed bat prefers warm climates and is most commonly found in northern South America, Central America and the Caribbean Islands. Several colonies discovered in building roosts in the Florida Keys are believed to be members of this family. This bat is about the same size as the Brazilian or Mexican free-tailed bat *(Tadarida brasiliensis). Molossus molossus* has protective hairs on the rump, but can be distinguished from other loose-tailed bats with protective hairs by its smaller size and wrinkled lips.

Wing expansion: 89-104 mm.

Weight: 10-14 g.

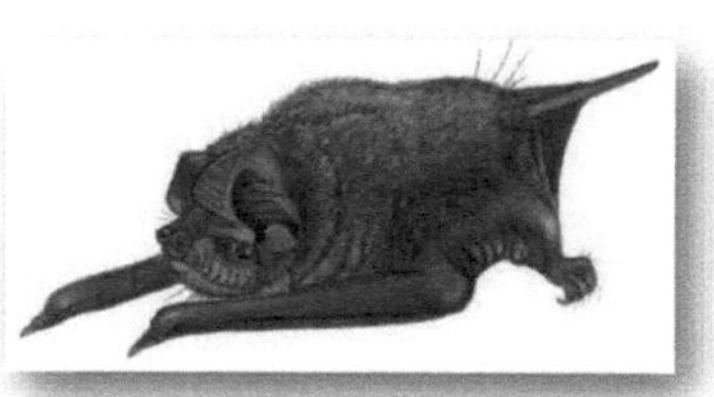

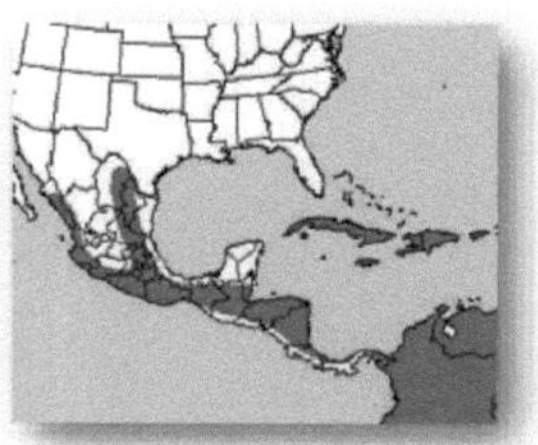

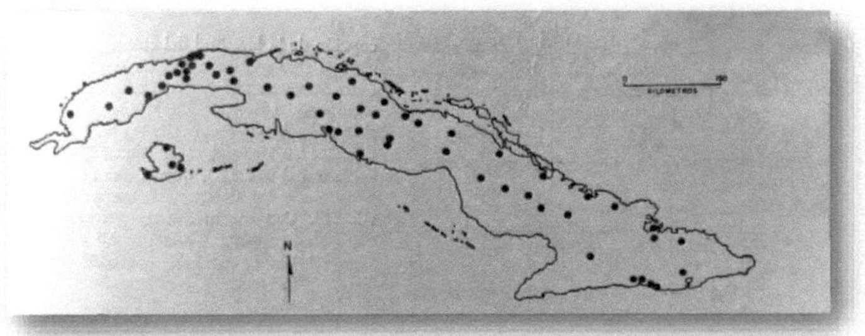

KILOMETROS
N

Bat with Western Biggest Bonnet

(Eumops perotis)

ORDER :Chiroptera

FAMILY :Molossidae

Description. Western greater hood bats live in steep rocky canyons typical of the arid regions of the southwestern United States, where they inhabit crevices in vertically sloping cliffs. Because of their relatively large body size and narrow wings, these bats cannot take off from a flat surface and must free-fall from a height to initiate flight. They hang head down in crevices and can detach, pick up speed as they fall and take flight with a flap of their wings to begin their nocturnal hunt for insects they choose as prey. If one of these bats is on the ground, it will climb a tree or other object to reach a sufficient height to launch itself into flight.

Also known as: Bat with Bonete Mayor, Murciélago con Bonete Mayor, Murciélago con Bonete.

Sexual Dimorphism: Males are larger than females.

Wing expansion: 159-187 mm

Weight: 45.5-73 g

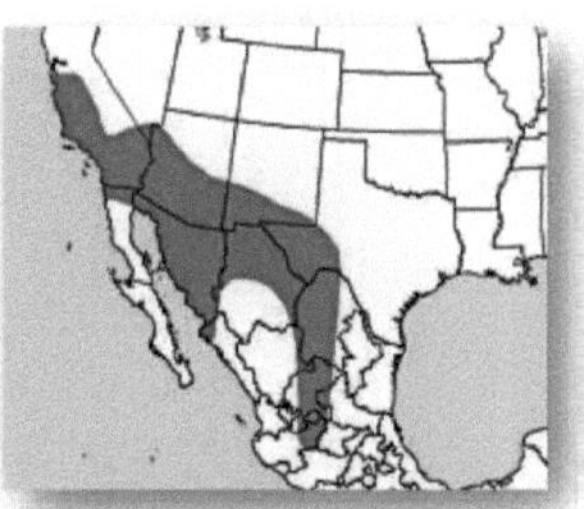

Eumops perotis - upper left (with *E. underwoodi).*

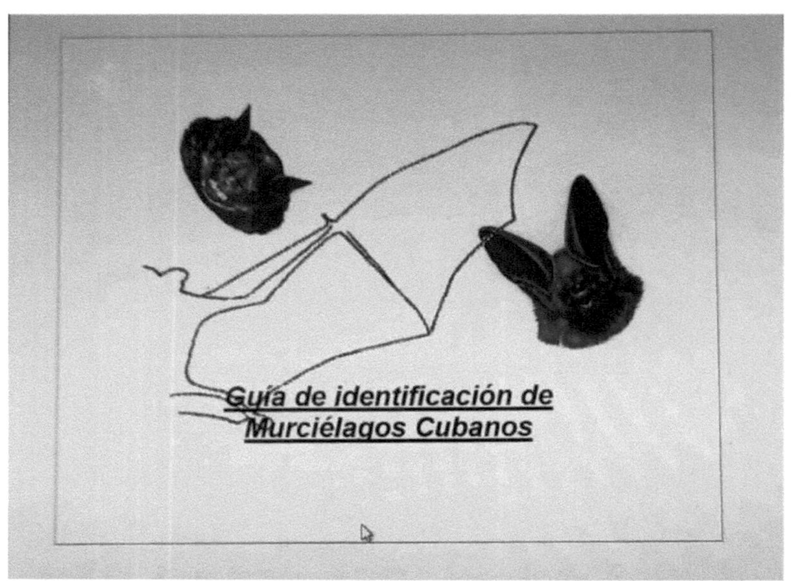

According to Gilberto Silva Taboada (1979).

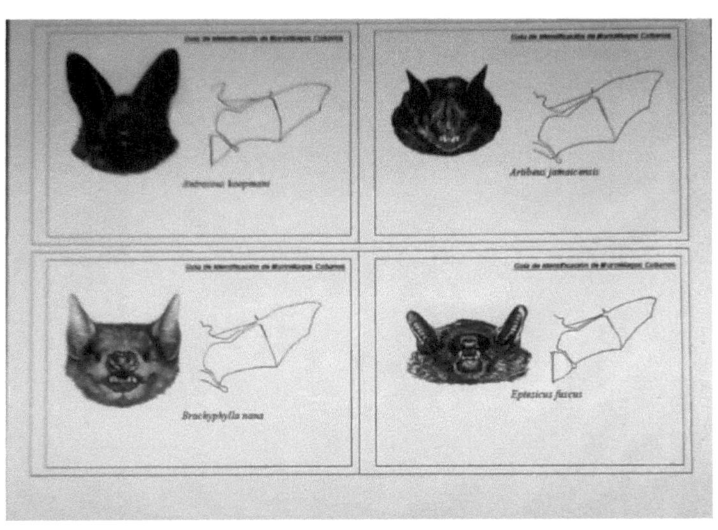

Guía de identificación de Murciélagos Cubanos
Antrozous koopmani
Guía de identificación de Murciélagos Cubanos
Artibeus jamaicensis
Guía de identificación de Murciélagos Cubanos
Brachyphylla nana
Guía de identificación de Murciélagos Cubanos
Eptesicus fuscus

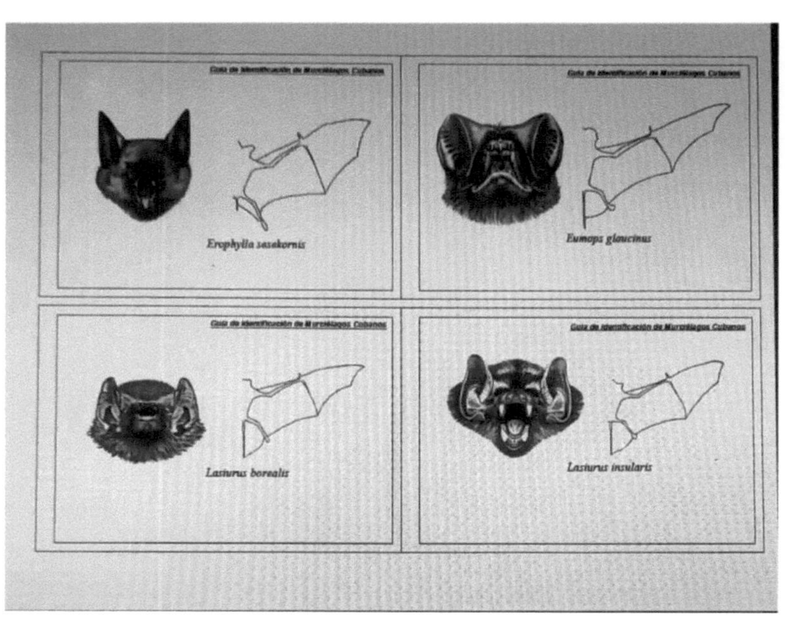
Guía de Identificación de Murciélagos Cubanos
Erophylla sezekorni
Guía de Identificación de Murciélagos Cubanos
Eumops glaucinus
Guía de Identificación de Murciélagos Cubanos
Lasiurus borealis
Guía de Identificación de Murciélagos Cubanos
Lasiurus insularis

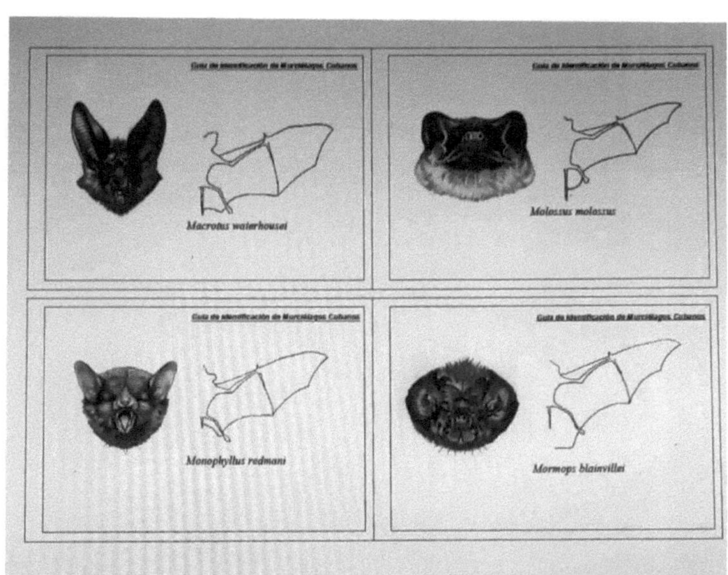

Guía de Identificación de Murciélagos Cubanos
Macrotus waterhousei
Guía de Identificación de Murciélagos Cubanos
Molossus molossus
Guía de Identificación de Murciélagos Cubanos
Monophyllus redmani
Guía de Identificación de Murciélagos Cubanos
Mormops blainvillei

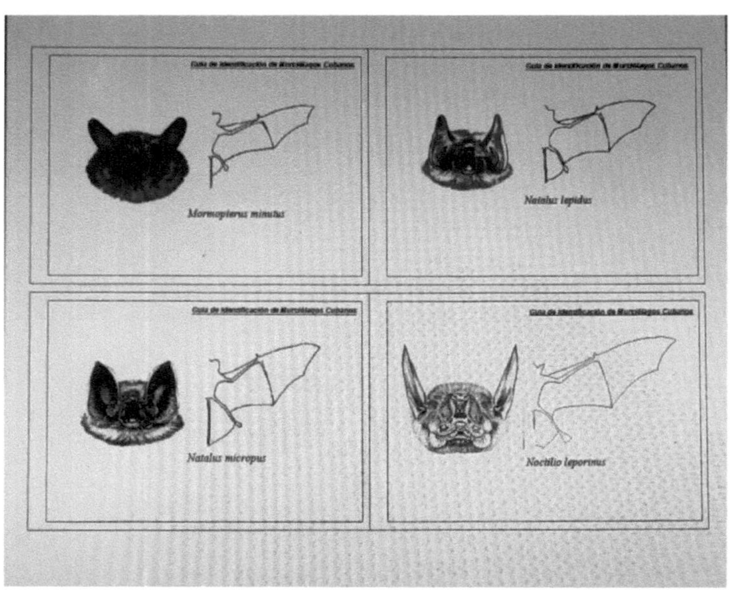
Guía de Identificación de Murciélagos Cubanos
Guía de Identificación de Murciélagos Cubanos
Mormopterus minutus
Natalus lepidus
Guía de Identificación de Murciélagos Cubanos
Guía de Identificación de Murciélagos Cubanos
Natalus micropus
Noctilio leporinus

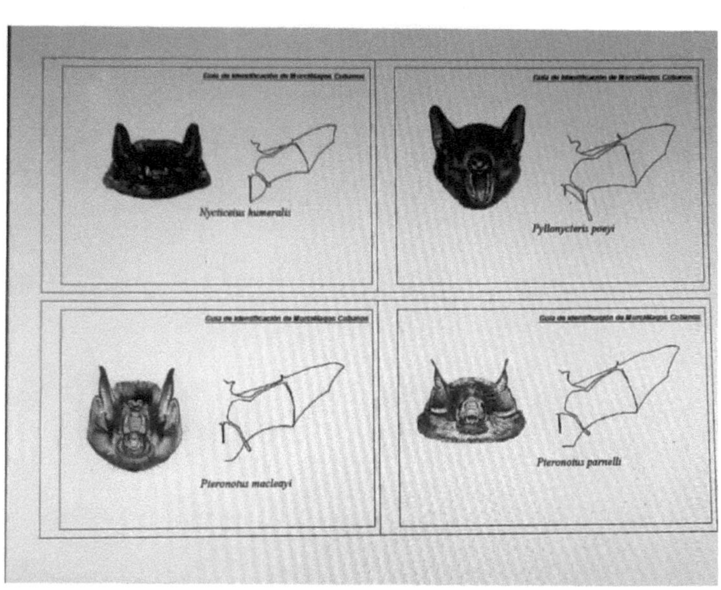

Guía de identificación de Murciélagos Cubanos
Nycticeius humeralis
Guía de identificación de Murciélagos Cubanos
Phyllonycteris poeyi
Guía de identificación de Murciélagos Cubanos
Pteronotus macleayi
Guía de identificación de Murciélagos Cubanos
Pteronotus parnelli

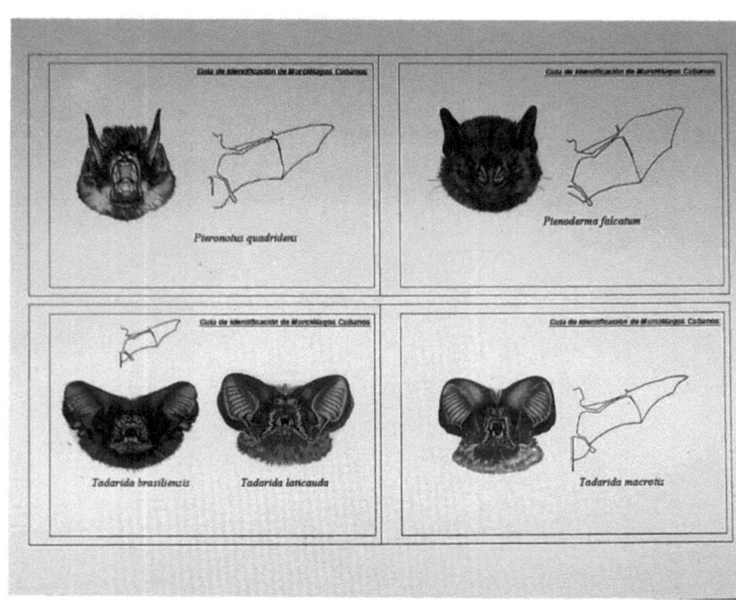
Guía de identificación de Murciélagos Cubanos
Guía de identificación de Murciélagos Cubanos
Pteronotus quadridens
Pteronoderma falcatum
Guía de identificación de Murciélagos Cubanos
Guía de identificación de Murciélagos Cubanos
Tadarida brasiliensis
Tadarida laticauda
Tadarida macrotis

According to Carlos A. Mancina.

MURCIÉLAGOS
de Cuba

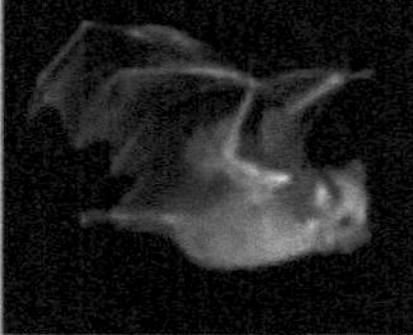

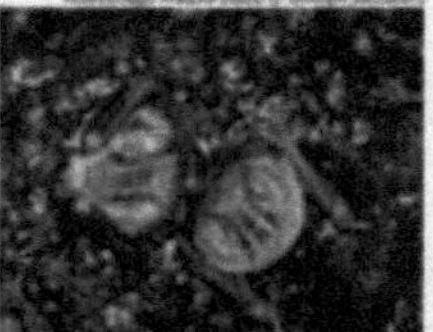

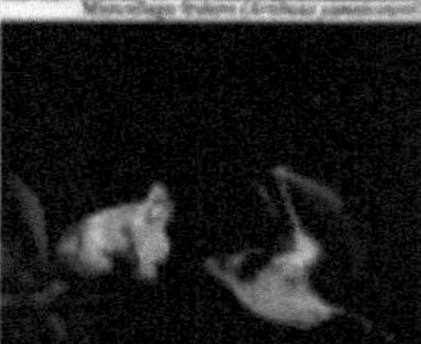

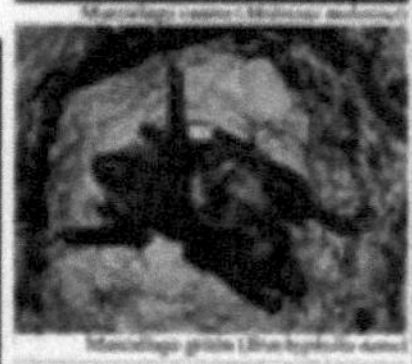

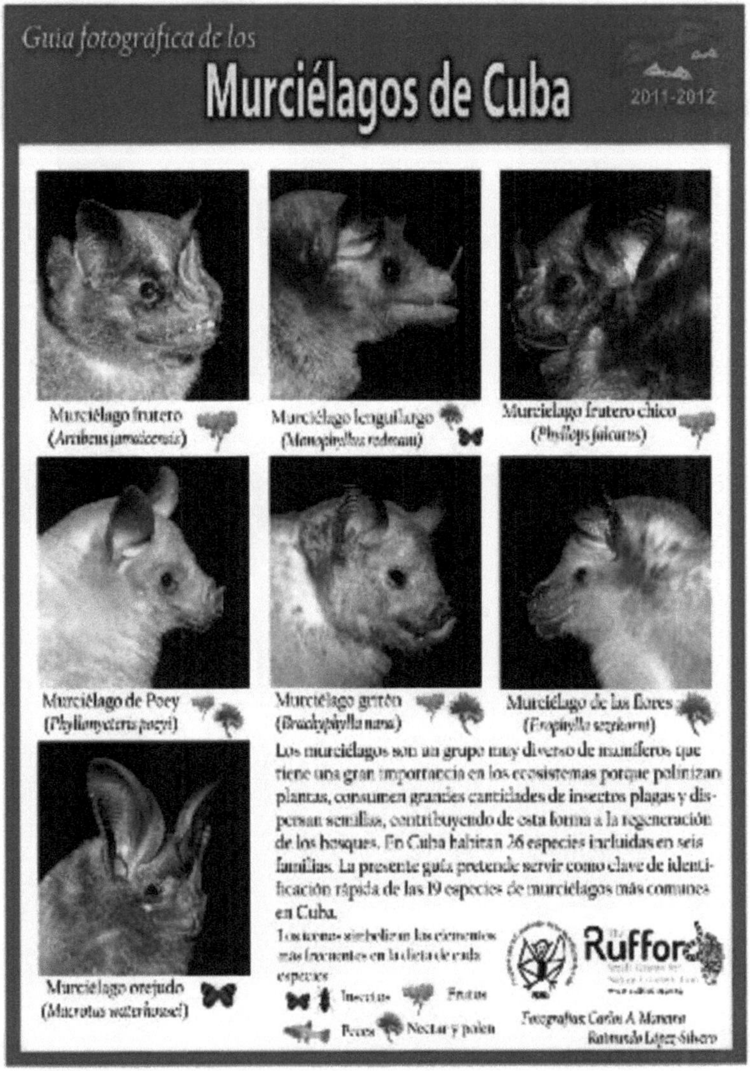

Guía fotográfica de los
Murciélagos de Cuba
2011-2012

Murciélago frutero
(Artibeus jamaicensis)

Murciélago lengüilargo
(Monophyllus redmani)

Murciélago frutero chico
(Phyllops falcatus)

Murciélago de Poey
(Phyllonycteris poeyi)

Murciélago gritón
(Brachyphylla nana)

Murciélago de las flores
(Erophylla sezekorni)

Murciélago orejudo
(Macrotus waterhousei)

Los murciélagos son un grupo muy diverso de mamíferos que tiene una gran importancia en los ecosistemas porque polinizan plantas, consumen grandes cantidades de insectos plagas y dispersan semillas, contribuyendo de esta forma a la regeneración de los bosques. En Cuba habitan 26 especies incluidas en seis familias. La presente guía pretende servir como clave de identificación rápida de las 19 especies de murciélagos más comunes en Cuba.

Los íconos simbolizan los elementos más frecuentes en la dieta de cada especie

Insectos
Frutas
Peces
Néctar y polen

Rufford

Fotografías: Carlos A. Mancina
Raimundo López-Silvero

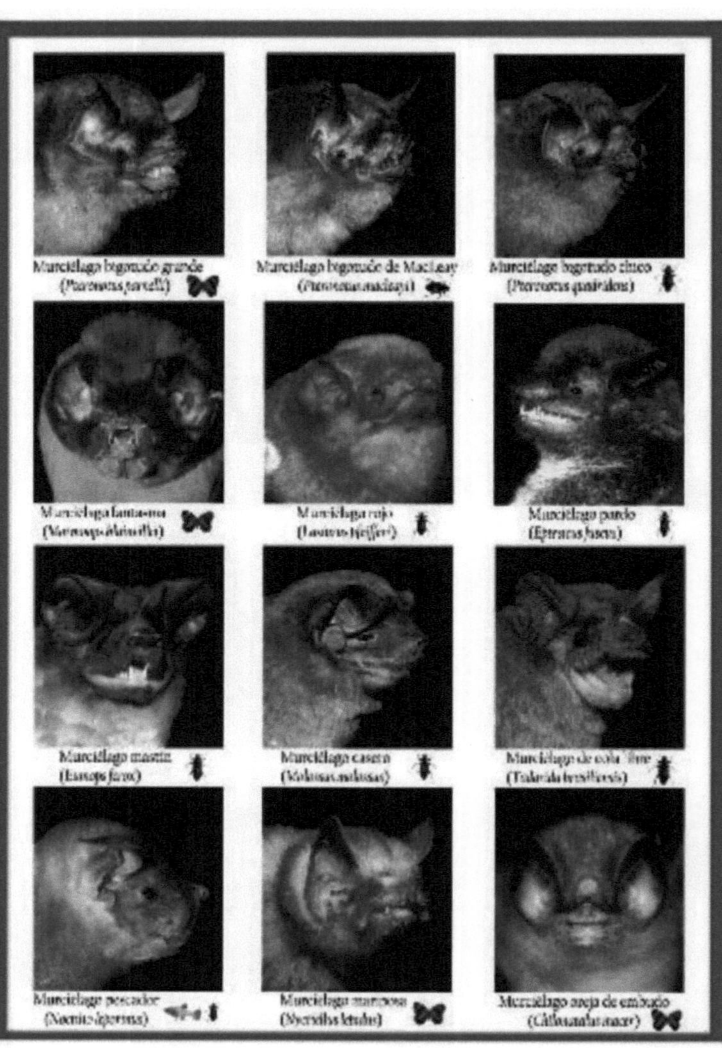

Murciélago bigotudo grande (Pteronotus parnellii)
Murciélago bigotudo de MacLeay (Pteronotus macleayii)
Murciélago bigotudo chico (Pteronotus quadridens)
Murciélago fantasma (Mormoops blainvillei)
Murciélago rojo (Lasiurus pfeifferi)
Murciélago pardo (Eptesicus fuscus)
Murciélago mastín (Eumops ferox)
Murciélago casero (Molossus molossus)
Murciélago de cola libre (Tadarida brasiliensis)
Murciélago pescador (Noctilio leporinus)
Murciélago mariposa (Nycticeius lepidus)
Murciélago oreja de embudo (Chilonatalus macer)

BIBLIOGRAPHY .

Arredondo, C.; R. Armiñana, N. Chirino y R. Agüero. 1996. *Zoología de los Cordados,* Volume II. Ed. Pueblo y Educación, Havana.

Borroto-Paéz, R. and C. A. Mancina. 2012. Mammals in Cuba. Fundación Espartaco- Sociedad Cubana de Zoología. Helsinsky.

Grahanm, G. L. (1994): *Golden Guide / Bats of the World.* Bat conservation International, Austin.

Harvey, M. J., J. Scott Altenbach & T. L. Best. 1999. *Bats of the United States.* Published by the Arkansas Game & Fish Commission in Cooperation with the Asheville Field office U.S. Fish and Wildlife Service.

Hernández-Muñoz, Abel. 2002. *Mammals that fly.* Editorial Gente Nueva. La Habana.

Hernández-Muñoz, Abel. 2013. *Duendes nocturnos.* Lulu Press INc. Los Angeles.

Hernández-Muñoz, Abel. 2015. *Bats, different mammals.* Lulu Press INc. Los Angeles.

Hernández-Muñoz, Abel. 2023. *The secret life of bats.* Editorial Académica Española. Schinof.

Silva Taboada, G. 1979. *The bats of Cuba.* Academia, Havana.

Tuttle, M. D. and D. L. Hensley. 1999. *The bat house builder's handbook.* Bat Conservation International, Austin.

Varona, L. S. 1980. *Mammals of Cuba.* Ed. Gente Nueva, Ciudad de la Habana.

yes
I want morebooks!

Buy your books fast and straightforward online - at one of world's fastest growing online book stores! Environmentally sound cue to Print-on-Demand technologies.

Buy your books online at
www.morebooks.shop

Kaufen Sie Ihre Bücher schnell und unkompliziert online – auf einer der am schnellsten wachsenden Buchhandelsplattformen weltweit! Dank Print-On-Demand umwelt- und ressourcenschonend produziert.

Bücher schneller online kaufen
www.morebooks.shop

Printed by Books on Demand GmbH, Norderstedt / Germany